Patrick the Paratrooper

By Chris Langlois, a Doc Roe grandson

Illustrated by Ricardo Lima

ISBN-13: 979-8-9874897-2-7

This book can be found on www.Amazon.com or by contacting the author at: docroegrandson@gmail.com

DEDICATED TO ALL THOSE, WHOM LIKE PATRICK, HEEDED THEIR NATION'S CALL.
ISAIAH 6:8

DECEMBER 7, 1941.

It was a regular Sunday afternoon. I was playing army, getting my troops ready for battle.

My name is Patrick and I'm the general of these toy soldiers.

Sure, we were against an enemy with more soldiers. But, we were better trained and most importantly, we stood in the fight together!

Then suddenly, my mom and dad yelled for me from the living room, "Patrick, come listen to the radio! The newsman said the Japanese have attacked Pearl Harbor!"

"The Japanese? Who are they?" I asked my parents. But they were too focused on the radio to reply.

Pearl Harbor? Where was that I wondered?

The newsman said Pearl Harbor was in Hawaii. Boy, that's a long way away from here. The Japanese airplanes had sunk many of our Navy's ships and over 2,000 sailors had been killed.

America was at war!

Like everyone else, I felt it was my duty to respond to our country being attacked. America was our home. The president said it was a date that no one would forget. Boy, was he right!

There was a rush to join the Army, Navy, Air Force and Marines. The lines of men went out the doors of the recruiting offices. I wanted to do my part, so I joined the Army. I raised my right hand and swore an oath to defend the Constitution of the United States.

On the wall of the recruiting office was a poster about the paratroopers.

I asked the officer in charge, "Sir, who are the paratroopers?" He said, "A paratrooper is an Army soldier who jumps out of airplanes with a parachute. When you land on the ground, you are already in battle, surrounded by the enemy. But, I'll tell you, the training is very difficult. And they only take the best of the best soldiers. Oh by the way, you have to volunteer."

I stood up straight and said, "Well that's me, Sir! Send me to be a paratrooper!!"

A few weeks later, I was sent to Camp Toccoa in Georgia for training. I had to become an Army soldier first before I could become a paratrooper.

I got off the train. It sure was hot in Georgia. And how could they have so many mosquitos?

I checked in with the sergeant who lined us in a row. He assigned us our tents to sleep in while we were in training. The sergeant also gave us our specific unit assignments...I was going to Easy Company.

Next to our camp was a tall mountain called Currahee. The sergeant told us that Currahee was an Indian word that meant, "Stands Alone." The sergeant reminded us that if we became paratroopers, we would parachute into enemy territory and that we too would, stand alone, until we could find our buddies. And even once we found our buddies, our whole group would still be surrounded. That was a pretty scary thought.

We had to run up and down the mountain almost every day...3 miles up, 3 miles down. That was tough!

But there was plenty more training we had to do too. We had to run an obstacle course, climb ropes and crawl through the mud. A lot of soldiers started the training, but only a few of us made it all the way through.

I guess they weren't joking about making sure we were the best of the best!

After Camp Toccoa, I had completed my training to be an Army soldier. Now I had more training to finish in order to become a paratrooper. I was going to be a part of a brand new unit, the 101st Airborne Division. The Screaming Eagle was the patch they wore on their sleeves. But first, I had to earn the right to wear that patch.

We learned parachuting is not as easy as just falling out of an airplane. We had to learn how the parachute works and how to land on the ground without hurting ourselves. We had to learn how to work together as a team once we landed.

After all the training, we had to make five parachute jumps in order to earn our jump wings. One of those jumps was in the dark of night. Everyone was both scared and excited at the same time. We wanted those silver wings on our uniform to show we were special soldiers!

After our five parachute jumps, we had earned our jump wings. They were pinned on our uniforms. Boy, we were proud!

After many months of hard work only the best of the best remained. Only a very few soldiers in the whole Army were paratroopers. We had trained hard together and sweated together – these were the best buddies a guy could have.

I was proud to be a part of Easy Company.

And now, I wasn't just Patrick. I was Patrick the Paratrooper!

But our days of constant training were coming to an end.

The war was being fought in many countries. The Marines and the Navy sailors were fighting the Japanese in the Pacific Ocean islands. But our destiny was leading us to free millions of people in Europe who had been captured by the Germans.

Easy Company sailed from New York City to England on a big ship and I took a photo with the Statue of Liberty as we left, to send to my parents. I know they worried about me going to war. But I was with a Company of brave and strong paratroopers.

My buddies and I were ready for battle.

JUNE 6, 1944. 1:00 AM.

They called it "D-Day." It was time for all of our training to be put to the test. We loaded our parachutes on our backs. We added food, ammunition and extra socks to our pockets and backpacks. We each had a shovel to dig foxholes, a canteen with water and our knives were strapped to our boots. Our gear was so heavy we had to help each other climb up into the airplane.

I could look out of the plane as we flew toward Normandy, France. There were hundreds of planes carrying thousands of paratroopers. Plus, there were hundreds of ships on the ocean that would bring thousands of more soldiers to land on the sandy beaches.

But the German army was ready and waiting for us and it was not going to be an easy fight. Their guns and cannons were firing from the ground up at our airplanes and some of our planes were hit and crashed into the ground with huge fireballs.

It looked like the 4th of July...except this wasn't a celebration.

We parachuted and landed in the dark...surrounded by the enemy.

There was gunfire and explosions all around us. It took us all night to find some buddies, to get into small groups and to reach our assigned location.

As the sun began to rise, we found four German cannons nearby that were firing on the beach where our soldiers were landing. Those were the same soldiers in all those ships I saw as we flew over, just before I parachuted. The explosions from those cannons were doing a lot of damage!

We gathered a group of Easy Company men and attacked those four cannons. We beat the Germans and destroyed the cannons. I know we helped save some lives of the soldiers on the beach when we silenced those guns.

We fought in Normandy for over a month. Some of my buddies were in the hospital, some of my buddies had been killed. But from fighting in combat together, we had formed a special bond.

We took ships back to England. Finally, we were able to get some rest.

New soldiers joined Easy Company to replace our injured buddies. We called them "replacements." We all trained together some more.

The Army gave us new uniforms and new supplies. We filled our backpacks and our pants pockets with food again. We were getting ready for the next mission.

Meanwhile, our Army had moved forward and pushed the Germans out of France. We were excited to hear the enemy was in retreat! We cheered on those soldiers still in battle.

And so, our next parachute jump was into The Netherlands. This time, we jumped during the daytime and there were fewer German cannons shooting into the air at our planes, and at us. Whew!

SEPTEMBER 17, 1944.

It was almost an easy jump and I let out a sigh of relief that I landed on some soft dirt.

But that easy jump did not last long.

We dug foxholes that filled with water, which means we were wet a lot of the time. We charged across a field with bayonets to attack an elite unit of German soldiers. We even paddled across the Rhine River in small boats, at night, to help rescue some British soldiers.

We had to fight the German army for 72 days in The Netherlands. More of my buddies were hurt and went to the hospital. And more of my buddies were killed.

But I did it all, side-by-side, with my buddies.

It was almost Christmas and we hoped the Germans would surrender and then we could all go home in time to open our presents under the Christmas tree. But that was not going to happen.

The Germans made a surprise attack and the 101st Airborne was sent in to help. But this time, we did not jump from airplanes, we jumped from the back of big, transport trucks. It was the coldest winter in a very long time and our boots hit the frozen ground in the small town of Bastogne, Belgium.

It was hard to dig a foxhole with our shovels as the ground was mostly ice. We did not have warm food. We did not have warm clothes. Besides the German bombs and bullets, the cold and the snow was just as awful an enemy, especially on our hands and our feet.

Oh, did I forget to mention we were surrounded by the enemy, again? But that's what paratroopers do! And we depended upon our buddies, even more than before, to get us through.

The German general asked for our surrender. Was he nuts?! We could not, and would not, give up!

When our buddies were injured, they screamed for the medic.

Sometimes, I helped carry the wounded men to the jeep where they could be taken to the hospital. Our medics were special soldiers. They did not carry a gun, they only carried bandages.

We called the medics "angels" because they just seemed to appear when you needed them. They took care of us all the time, especially when you got wounded. All the soldiers respected the medics and we looked out for them, especially when the battle began.

After Christmas, we beat the Germans in Bastogne. We continued fighting, pushing forward into Germany. We finally reached the Eagle's Nest high in the mountains. The views were amazing.

The fighting was over. It was the end of the war for us. It had been a long and tough war.

We were able to stop being soldiers for a while and even got a small reminder of home with some time to play baseball.

As I sat and enjoyed the beautiful scenery of the mountains with the blue skies above and the sun shining, I thought about my buddies. I thought about my buddies who were still with me, but more importantly, I thought about my buddies who were still in the hospital; some of them had lost their legs in the battle of Bastogne.

And I thought about my buddies who had died. They would not be able to see their parents again. And boy, did I miss my parents. I'm one of the lucky ones.

But my buddies were more than just buddies now. They were my brothers and we had become a Band of Brothers. I'll never forget them. I know I'll think about them every day...for the rest of my life.

PRESENT DAY.

The light grew from a sliver and slowly began filling the room that was just a second ago, quiet and dark.

"Patrick, c'mon sleepy head, you're going to miss the school bus. I have your breakfast ready." Mom said as she opened Patrick's bedroom door.

Patrick popped up from under the covers of his bed.

"Mom, I had the most amazing dream!! I was a paratrooper in World War 2 with Easy Company! And then I parachuted out of an airplane on D-Day! And then, my buddies and I..."

Chris Langlois is a grandson of medic Eugene Gilbert Roe, Sr. "Doc" Roe joined Easy Company at Camp Mackall, just after Camp Toccoa. Chris is originally from Baton Rouge, Louisiana and graduated from Louisiana State University. He currently resides in Dallas, Texas with his daughter, Julia. Chris is a police officer on the streets in Dallas, Texas where he enjoys meeting new people and placing them in jail.

For older students and adults, Chris' first book, *How Easy Company Became a Band of Brothers* also follows Easy Company through text and beautiful illustrations by Anneke Helleman. Find more about Doc Roe Publishing on Facebook, Instagram and Twitter. Chris can be reached at: docroegrandson@gmail.com

♠ ☂ ♠ ☂

Las Vegas based illustrator, Ricardo Lima, honed his skills at the Academy of Art University before becoming a freelance illustrator taking on projects both large and small. From an early age his love of comics led him to become passionate about the craft and with training in both Traditional and Digital Art has defined his art style. Somewhere between realistic and cartoony, his bold shapes and steady linework has given him the opportunity to express his creativity in different artistic worlds. Ricardo can be reached at: www.ricartolima.com

www.ingramcontent.com/pod-product-compliance
Lightning Source LLC
Chambersburg PA
CBRC092313210326
41597CB00043BA/165